More Praise for *Some Aid to Navigation*

Carey Taylor's work is steeped in the wild reaches of the Northwest coast—crow-dark forests, seaweed tang of low tide, the boom of Pacific surf. Like lush kelp ribbon anchored on jagged rocks, beneath these tender memories lies tumult and loss. *In the beginning/ we knew nothing / of charred ground.* These poems braid a layered journey in finding new bearings that will *meld our/ fractured parts into some kind of beauty.*

Gina Hietpas, author of *Terrain*—
Finalist for Washington State Book Awards

Some Aid to Navigation

Some Aid to Navigation

CAREY TAYLOR

MoonPath Press

Copyright © 2024 Carey Taylor.
All rights reserved.

No part of this publication may be reproduced,
distributed, or transmitted in any form or by any means
whatsoever without written permission from the publisher,
except in the case of brief excerpts for critical reviews and articles.
All inquiries should be addressed to MoonPath Press.

Poetry
ISBN 979-8-9899487-1-0

Cover art: Mary O'Shaughnessy
Portal to Time and Tide
9 x 12 letterpress, monoprint & colored pencil
www.marykoshaughnessy.com

Author photo: Sean Taylor

Book design: Tonya Namura, using
Adobe Jenson Pro (text) and Colaborate Thin (display).

MoonPath Press, an imprint of Concrete Wolf Poetry Series, is
dedicated to publishing the finest poets living in the U.S. Pacific
Northwest.

MoonPath Press
c/o Concrete Wolf
PO Box 2220
Newport, OR 97365-0163

MoonPathPress@gmail.com

https://MoonPathPress.com

For my parents

The cradle rocks above an abyss, and common sense tells us that our existence is but a brief crack of light between two eternities of darkness.

—Vladimir Nabokov

Contents

I

Birth Announcement	5
The Days of Starfish	6
Inheritance	8
Ode to Lust	9
The Coxswain	10
Wapiti Blues	11
Ways to Save a Life	12
Oregon's Intertidal Zone Approaches Tipping Point	13
We're Heading into Nut Country Today	14
The Bambino Year	15
Western Pleasure	21
Suicide in a Small Town	22
War Play	24
Endangered	26
Photograph of My Parents—Port Orford 1955	27

II

Returning	31
Universe Expanding	33
My Father Leaves on Mother's Day	34
The Family Subdivides	36
Family Dog	38
Wreckage	40
They Never Talk to Each Other Again	41

Solastalgia   42

Closing Time   43

Early Seral Stage   44

Between the Past and the Future   47

Ode to Noxious Weeds   49

Extinction Dreams after Beachie Creek Fire   51

The Lighthouse Keeper's Daughter   53

Driving over the Columbia River   55

Lifeblood   56

Tide Change   57

Some Aid to Navigation   58

III

Siren Song   61

The Lighthouse Keeper's Daughter Calms Herself Down   62

Surfacing   63

Ode to My Belly   65

Garibaldi   66

Ode to Being Fed   67

Where God Resides   69

Sometimes a Meteorite Becomes a Moon   71

Praise the Childhood of Borderless Roaming   73

Ode to the Cooper's Hawk   75

Towards Kintsugi   76

It Came to This   77

Birthday Fires   79

The Lighthouse Keeper's Daughter Goes Home   80

Notes	81
Acknowledgements	83
Gratitude	85
About the Author	87

Some Aid to Navigation

I

*They went to sea in a Sieve, they did,
In a Sieve they went to sea…*

—Edward Lear

Birth Announcement

Let's begin with my feet. Ten proper toes
pressed into pink ink

then pressed again on one
side of a small card—two tender rhodies

determined to root
in the heavy dark of a Port Orford winter.

The first question my father asks the nurse
is not *is it a boy or girl,*

but *is everything ok?* When she answers yes,
his fear unwinds like coiled rope,

after drop of anchor. As he drives from hospital
to home, he looks at my face in the cardboard

box between himself and my mother. He thinks
I look like his sister Georgie,

whose old crib he has repainted and is the newest
piece of furniture he owns.

At twenty-two, what he doesn't know yet,
is that the 36-footer moored at the boathouse

in Nellie's Cove has sprung a slow leak.
What he doesn't know yet,

is that he will try every lifesaving trick he has learned,
to keep both girl and boat afloat.

The Days of Starfish

My lungs fill first with cold
Pacific air.

Mother sings to me.
Father fishes the kelp beds.

Before I walk
I reach for starfish

that cling
to tide pool rock.

Orange and purple
stellar bursts

bejewel
steel days.

*

It starts with a common ocean virus,
a sea beginning to warm,

a simple immune system
unaccustomed to heat.

The first signs are skin lesions.
Then twisted arms, a body

that deflates, then liquifies
into a puddle.

When starfish no longer
eat urchins, urchins multiply,

clear-cut forests of kelp.
Kelp beds shelter.

Sequester carbon.
Feed other species.

*

I build driftwood
forts

on our front yard
beach.

Dig razor clams
by hand.

Stand barefoot
at the waterline.

Watch my father
cross the bar.

Starfish. Father.
Father. Starfish.

Inheritance

My father came
from a long line

of drifters
who craved pink.

Eden roses
on weathered arbors.

Fillet of salmon.
Wind-whipped cheek.

Women with toenails
the color of carnations.

While I, like the women
before me,

waited
in the chicken yard,

for the crunch of gravel
on the road,

the tired hello,
the ocean in their kiss.

Ode to Lust

She looked steamy
with that highball
in her hand
black dress fringed
in red crinoline
cigarette ready
for any overflowing
ashtray
smoldering
around the room.
Of course he
wanted her.
Like he wanted
a paper boat as a kid,
maybe the old Buick,
maybe as needle and thread
in the gravitational divide
between coming and going.
He wanted to bury his face
in that immaculate shingle of hair
at the back of her neck,
to inch his way
toward jaw and ear,
static fuzz of temple.

The Coxswain

My father moves a three-legged table
from under the east window to my bedside.

The ceramic piggy bank I got for my seventh
birthday falls to the floor and shatters.

He plugs in a humidifier, turns the knob. Steam
mists into the dim room. He tucks a wool blanket

with a tattered silky edge around my shoulders,
pulls a chair beside my bed and sings "Waltzing Matilda."

When my wheezy lungs
are no longer a hive of bees stinging,

and my breath calms,
he lifts his palm from my forehead.

When I wake in the morning, he has crossed the Tillamook
Bar—headed west into the vast Pacific.

My god, I realize, all my childhood he carried my shallow
breath over those wind-whipped waves while navigating

dangerous waters to save others. All those years
when I thought my drowning lungs were my own,

he was bent over with outstretched arms,
throwing me a life ring.

Wapiti Blues

Wapiti is from the Shawnee and Cree term for elk

In this picture, my father wears a plaid
Pendleton shirt, a cigar hangs from his mouth. Standing
beside him, my cousin grips a six-pack of Olympia.

Outside the frame, I sit on a downed log,
my father's rifle heavy on my lap.

I remember her massive body hauled in pieces
from the bottom of the ravine, her head propped up
against two others in the bed of my uncle's pickup truck.

I remember her tanned hide was made into soft chaps and
gloves, the freezer crammed with steak and sausage.

I remember a big storm that took the power out
for days, and a bad smell wafting up the
basement stairs.

I remember my mother scrubbed and scrubbed for weeks,
but the scent of corpse and bleach persisted.

Ways to Save a Life

After my father rescues three doctors whose boat capsizes
off the Tillamook Bar, I fidget in a new dress, wait for
the cake to be cut, beg to stand next to him while he is
honored with a Coast Guard medal. At seven, I do not fully
understand how my father saves lives, but when I am older
he tells me he rescues mostly fools and bad-luck fishermen,
sometimes both. Pulls them off sinking boats. Hauls them
hypothermic from the choppy Pacific. Tows their derelict
skiffs to safe harbors. Lifts broken bodies from jagged rock
to top of cliff in a Stokes basket.
      But of course, you can't save everyone.
Deckhand who stumbles overboard on a moonless night
after his fifth Rainier. Bridge-jump-bone-shattered pool of
the old widower now bagged on the bow. Teenagers pulled
into the galaxy currents of Rosario Strait. And even a sailor
trained to save lives can panic—forget the life ring hanging
from the rail below the cockpit, forget the puffy orange
jackets in the hold, forget the time he circled the dark green
to rescue his overboard crew, forget his own girl treading
in his icy wake—yes, even a sailor trained to save lives can
panic—change his heading south toward winds of clove and
pepper, profusion of bougainvillea on bone-white arbors.

Oregon's Intertidal Zone Approaches Tipping Point

He sang in the shower. He sang in the car.
He sang on the dewy lawn scouting for stars.

> The song on the dewy lawn still makes me cry.
> *I'm sad to say I'm on my way.*

His sad Belafonte songs spun away from home.
Won't be back for many a day.

> Can we ever come back after so many days?
> My father's voice was both anchor and fortune-teller.

My father pulled anchor and took his voice.
Once at Turn Point Light we ate mussels from small pools.

> Tide pools filled with mussels we didn't know were
> leaving us, as he sang *cockles and mussels alive, alive-o.*

He fed me cockles and mussels and "Molly Malone."
He sang in the shower. He sang in the car.

We're Heading into Nut Country Today

After the announcement over the intercom
our third-grade class goes silent.

Mrs. Porter stands up from her desk,
her face the color of powdered milk.

She tells us to grab our coats and lunchboxes,
ushers us onto early buses home.

My mother is not waiting for me with celery sticks
and Cheez Whiz or her frosted ginger cookies.

She stands in front of our black-and-white TV,
hair still in curlers, hands clasped around a tea towel.

For the rest of our lives
we talked about Jackie in her bloodied pink suit,

scrambling on hands and knees
on the back of that dark limo.

How after being pushed back into her seat,
she cradled his head on her cherry-blossom lap.

The Bambino Year

In a plastic bag
tucked away in the attic

I find ten postcards
from my dad

the year he lived
in Sellia Marina.

Hi Pumpkin!
How's my gal?

I love you.
Can't wait to be home.

I can't find the
doll-purse he sent me

filled with
hard Italian candies

or the blue and white
seersucker dress

with kittens
embroidered on the pockets

but I remember sashaying
to the front of my

third-grade class
for show-and-tell—

purse handle wrapped
around my wrist

dress hem stopping
just above the knee

proclaiming to the
entire room

I was someone's
Italian girl.

*

A black-and-white
photo

of my dad
and Father Bruno

fall from an
airmail envelope.

They are standing
in a receiving line

in front
of a church

somewhere in
Calabria.

One wears the black
uniform of a cleric

one the Dress Blues
of the Coast Guard.

Every face is somber
after having passed

the empty casket draped
with the American flag

and its wreath of flowers
holding the framed

photo
of John F. Kennedy.

*

We all became
Catholics.

My dad in
Catanzaro.

Me, at St. Mary's
in Florence—not Firenze.

My mom is already
Catholic

from a long line
of French Catholics

who emigrated
from Normandy,

to Quebec,
to Wisconsin,

and now here—
to a small town

on the
Oregon Coast.

We practice
our faith for a year,

maybe two.
Later, I will

send my own
children

to Catholic
schools.

I want them
to memorize

Hail Mary and
Our Father.

I want them
to have someone

they can
call on

after they
board a plane.

*

The year
my dad is

transferred to Italy
to a Loran station

my mother takes
pictures of herself.

Sexy pictures of
her in nothing

but the sweater
she knitted for him.

Provocative pictures in pillbox
hats and pointed heels.

Pictures she slips in
with her letters.

I didn't notice
it then

but she
was beautiful.

*

I stand outside on
an observation

deck with my aunts,
uncles, and cousins.

I am holding
the hand

of my mom.
When my uncle

points to a plane
beginning to land

and says it's my
dad's plane

home from
Italy

all I see
is black exhaust.

I am afraid
the plane is on fire

but my uncle
just laughs and

says no. When
I meet my dad

inside the airport
I begin to cry.

When he bends down
and picks me up

I hold tight
around his neck.

Western Pleasure

My dad walks out the front door onto the porch
of the Coos Bay Lifeboat Station.

He is thirty-three years old and handsome in his uniform.
He takes a few steps down the stairs and sits.

The air smells of freshly mown grass
and low-tide seaweed.

I am twelve and sit tall in the saddle on my new horse. I am
wearing suede cowboy boots and buttery yellow chaps.

Dad is teaching me what to expect when I compete
in my first show.

First, I practice figure eights. Then he calls out
for me to walk, lope, trot, reverse direction, and back up.

We practice until he stands and says *good job Kiwi*—my cue
to ride the wooded trail back to the barn.

It was such a small moment.
His voice so tender.

Something slipped beneath my skin.
Something slumbered in my marrow.

Suicide in a Small Town

I can't remember your name
or even the grade you were in,

though I watched you step off
the bus every day after school.

I remember your unpainted house,
how the front stoop listed toward

the stagnant lagoon, how your yard
was a patch of sand and oyster shell.

I never asked you to play, and you
never asked me,

as if there existed some invisible line
we were afraid to cross.

And when someone, I don't remember
who, asked if anyone in our sixth-grade

class needed a ride to your funeral, I surprised
even myself when I raised my hand.

I hardly knew you, but I took
to your death like a moth to a porch light,

drawn toward your pale, still hands,
your small blue suit.

And even though I was young,
I sensed that under your tightly buttoned

shirt, a wheel of night even darker
than death had stopped turning—

and you were just waiting
for the cool ocean

currents
to carry your offering home.

War Play

> *Well boys, I reckon this is it. Nuclear combat,*
> *toe-to-toe with the Russkies.*
> —from *Dr. Strangelove*

Suddenly it is 1966 and I am playing with the
neighborhood kids in the front yard in Charleston, Oregon.

Usually, I am on the beach where the wind tangles my hair
in knots. Usually, I am riding my horse through the crow-

dark forest. Sometimes, I ride my bike past a chainlink
fence that surrounds the naval base, where someone listens

for sounds of Soviet subs under all that ocean. Sometimes, I
pick up peacock feathers in my neighbors' pasture,

or practice duck-and-cover at school. Always, I hear the
clanging of the American flag as it is raised and lowered on

the white metal pole outside my bedroom window.
But this day, I am hiding behind the west corner of our

large white house, nestled between the cedar siding and a
wild fuchsia.

We rock-paper-scissors to see who is the enemy. Bored
from so much waiting, I spray water on my boots from a

plastic pistol, suck nectar from the fuchsia's pink and
purple flowers, keep an eye on the boy behind the trunk

of a tilting pine. When I make my move, he screams to the soldiers on his side,

The Russians are coming! The Russians are coming!

Endangered

The orcas off Burrows Island were a pull greater
than her mother in the kitchen

or her father at the boathouse. Greater than sugar-frosted
flakes or fried eggs

or sitting on the outside stoop
with her yellow dog.

Greater than her brother's freckled cheeks
in the bassinet, or Captain Kangaroo

on the black-and-white television. More urgent
than her rain jacket hanging from the doorknob

in the mudroom.
It moved her feet down the narrow sidewalk,

past jangle of flagpole, the wire-brushed lighthouse
waiting for paint. On the edge of the cliff she gazed

beyond the jagged rocks below her, charmed
by their squeaky whistles and chirps, the glisten

of their black and white bodies rising
from the swirl and whoosh of kelp.

Photograph of My Parents—Port Orford 1955

They are standing
in front of a small white house.

She faces the camera
smiling.

Her left hip is hidden
behind his waist,

her left arm
slips under his armpit

and up his back,
until her hand

playfully
covers his eyes.

His right arm wraps
around her neck,

drapes
off her shoulder,

his wrist held in place
by her other hand.

His hair is shiny-dark.
Military short.

She sports a blonde
pixie cut, wears

a short-sleeved top
and pencil skirt.

He wears a chambray shirt,
sleeves rolled

above his elbows,
paint-splattered dungarees.

Both of them are wearing
their wedding bands.

They seem affectionate,
a little silly,

maybe
even happy.

O loves,
I have no words yet,

but even if I had, I doubt
I would have warned you.

II

Solastalgia:

The homesickness you have when you are still at home because your lived environment is changing in ways you find distressing.

—Glenn Albrecht

Returning

There is a place in me
that still knows water.

Where the gusts of time and
wind have etched me.

Where language began—clam,
cod, barnacle, bulb of kelp.

*

In a photo, I sit on the grass with my
mother and grandmother—

lighthouse behind us.
My father may have taken

this picture, or he may have
been tending the light.

He was an expert
at polishing the lens.

In either case, he is preparing us
for his absence.

*

On the bridge between mainland and island
I am the shadow on the plank.

Light of Cape Arago that no
longer circles back.

Mussels cracked and buried.
Midden waiting for a future dig.

Universe Expanding

I slip a gold hoop in each healed wound,
notice the right hole is glacier close to calving.

Has it always been like this? Or has gravity
pulled it down like the rest of me?

Did I not numb it enough with that fast-melting
ice? Or did I flinch when my father pushed

a warm needle through that fleshy lobe? Maybe
my mother slipped her hold on that tea-towel-

wrapped potato, as we sat side by side on that
faux leather sofa in our small tract house.

I remember my mother smelled of musky
perfume, her white legs in a miniskirt,

how my father's black hair was an inch or two
beyond Coast Guard regulation.

It was 1969—men on the Moon, boys in jungles,
hippies in beads, and our small triptych—

redshifting our way into the Age of Aquarius.

My Father Leaves on Mother's Day

I remember best
the call at work,

tossing my apron
into a lettuce box,

driving past the moldy
green single-wide

next to the creek,
the rusted cars and chickens

in the Crawfords' yard,
Scotch broom

the color of egg yolk
bordering our driveway,

the soft puddle
of my mother in bed.

I remember somehow loading
her into the Chevy Vega,

driving into the city on
unfamiliar streets,

walking into the emergency
room of an unfamiliar hospital,

the heavy door, a word
above the door: *Psychiatric*.

I remember best floating
back past field and fence.

The quiet, dark house.
Calling for the outside dog—

taking his flea-ridden body
into bed with me.

The Family Subdivides

By October
I realize

there is no way
back.

My father sits
in a studio apartment

and chain-smokes.
My mother

drifts like a
ghost ship.

The teacher next door
invites me over.

The horses stand
in knee-high grass.

The dog huddles
in the furnace room.

The apples won't
become pies.

I move forward
though not ahead.

It will be years
before I listen

to the hum
in my chest

become a rattle
I will pound at.

Before my hands
make tender

the tough
steak of me.

Family Dog

After her husband
and children move out,

she stands in a room
smelling of urine and bleach.

She strokes his dull
brown ears,

the spot
on his back,

his warm
freckled belly,

then lifts him off
the metal table,

wrapped
in a threadbare towel.

She lays him
on the passenger seat,

drives ten curved miles
past farm and fir,

carries his cooling
body

across
the back pasture,

digs hard ground
at the edge of the ravine—

blankets his body
with rock and clay.

Wreckage

After the teacher finished his grading.
After he flipped his porch light on.
After the marijuana brownies.
After listening to *Goodbye Yellow Brick Road*.
After the room started spinning.
After he carried me to his bedroom.
After he undressed me.
After I stared at the popcorn ceiling.
After no goodbye kiss.
After walking the dark driveway in my floppy black hat.
After the dog refused to leave its house.
After I stumbled through the back door.
After my mother was waiting for me.
After I took a shower until the water cooled.
After I crawled into my childhood bed.
After my mother tucked me in.
After my bedroom door softly closed.
After I lay awake for hours.
I quietly rose and bathed again.

They Never Talk to Each Other Again

On the Italian
blue plate

of their history,
they became

Trippa alla Romana—

soft belly of cow
with tomato,

Romano,
and mint.

Peasant
food

the upper
classes

would not
eat.

Solastalgia

pacific northwest bakes // starbursts this fine
bluebird day // sometimes violence isn't bruise
or blood // sometimes it's a whale pushing her
dead calf through once-cool waters // sometimes it's
the droop of dahlias in our grandmother's
garden // missing the scurry of earwigs in our
summer bouquets // red moon // red sky // red sun //
and nowhere to run // sometimes it's a teenager
chucking fireworks down the gorge // or smoky ash
on playgrounds where children cough red rover //
sometimes it's an omega block broiling
our bones // combusting on the shore of our
careless days // do you remember // once // twice
beneath damp cedars // long leash of seal in
sun-green kelp // incoming tide on our tongues

Closing Time

The bartender wipes the sticky table with a gray-white
rag, hawk-eyes the place for jacked-up voices, guy who

never smiles, pistol on the hip. He can feel it. This edgy
world, this live-for-the-moment world.

As the news streams a constant loop of disaster above
the shuffleboard, he needs a new drink to usher in summer.

He wants three months of effervescence—a sparkling
cocktail that goes down easy.

He mixes prosecco, elderflower liqueur, splash of club
soda, drops in three ice cubes, a twist of orange peel.

He lists it on the booze-board as the summer
special: St-Germain Spritz.

He wants to tip his glass to France, or Florida before sunlit
days of intolerable heat.

He wants to toast a cool breeze off some jade-green
ocean, the half-empty bottle of bourbon slow-loping

in his mouth, the flicker of fly in the fruit trees,
feather of the great blue heron.

Early Seral Stage

We rode together,
you and I,

down that blanketed
coastal road.

Port Orford to Bandon
and back again.

You at the front of the bus
so as not to get queasy,

me, inside the ocean
of your belly.

In the
beginning,

we knew nothing
of charred ground.

What came before.
What would come after.

But after
losing my father

and now you,
I track back

to the same ground
we bumped along

so many
years ago.

*

I discover Bandon
burned to the ground before

I was born.
Realize not all fires

can be contained,
nor are all fires lethal.

I stand outside
the hospital

where you delivered
the pomegranate burn of me.

See a girl
inside a girl.

*

I know now,
words were

door slams
that hurt me.

Sometimes
red taillights.

Always
the smell of horse.

I see how hard
I tried to wet our house

from ember.
Control the burn.

*

O loves, I may
forgive you.

I may not.
Either way,

there was
fireweed,

vine maple,
soft green

of huckleberry,
the purple bell

of
foxglove,

where
hummingbird

sugared
its forked tongue.

Between the Past and the Future

He often begins our phone calls
Hey Kiwi—

it's long kee and long wee,
nickname, anchor, sweet black licorice.

Someday, his sailor sounds will no
longer travel wire or wave.

Someday, I will wander the
rooms of his house without him.

I will look at his library of books, his
framed military medals, the model of

the 52-footer. I will smell him
in his closet full of bespoke suits.

I will pick up the symphony tickets
on the kitchen counter and remember

how we raced from the concert hall
to the ferry at the end of a show.

I will stare at the picture on his desk
of his father sitting on a motorcycle.

I will marvel at the quantity
of silk flowers he bought at Home Goods,

then placed in vases large and small
for every room in the house.

I will search for the gold necklace given
as a gift from his friends in Italy,

the one with the face of Jesus on one side
and a cross and anchor on the other.

I will spot the deer in his garden,
rush into the garage,

slam my stockinged feet
into his walk-to-the-mailbox clogs.

I will race over his well-loved lawn
with arms flailing, and scream—

Leave those goddamn roses alone!
And somehow he will hear me.

He will say as he did when I was ten,
Come on gal, let's take a picture. He will

shut the tattered curtain in the photo
booth at Pony Village Mall, and we

will sit on a small seat as the camera
snaps four or five black-and-white pictures,

which are delivered to our hands
in a single strip.

Our eyes will be bright, and with
his arm around my shoulder,

I will lean into the ocean
always lingering on his jacket.

Ode to Noxious Weeds

After it's official,
July was Earth's

hottest month
on record (again),

I buy French-lace
marigolds, push-up bra

yellow,
to entice bees.

I want   *oh, honey*
buzzing in my ear,

sweet clover
on my chin,

and I want it
again and again.

So I stuff in
one more purple petunia,

invasive ivy
I believe I can contain,

fill my flowerpot
to the verge of

collapse—as if the darlings
I tried to poison

in the alley
aren't hanging over

my weathered
fence

wilding
green and lush,

aren't sprouting fine
fringe

on their blackberry
buttons,

aren't eagerly
waiting

to be called
Rose.

Extinction Dreams after Beachie Creek Fire

In that coal-mine sleep before dawn,
    before shell of light opens, you find yourself
        at a party with Brad Pitt.

Drunk and looking sad, he places his hand
    on the nape of your neck, fingers your
        fishing-line scar, leans in and kisses you.

He's a smoky kisser and a sloppy drunk,
    but in this dream you are younger,
        so you say nothing.

Next you are with your realtor, looking for a new
    house. The well has run dry. Water is trucked in
        as they once delivered heating oil.

A new electric fence barricades the property to shock
    feral pigs. No blade of spring in the yard. Rain
        boots ache for scent of daphne.

Suddenly you are picking apples in your dead father's
    orchard. From a ladder he hands you Gravenstein after
        Gravenstein. You fill buckets for pies.

From his perch he looks north to Canada.
    Declares smoke-filled skies
        have cleared.

Awake, you grab your phone and check the air quality.
    Particulates are low. You throw open doors and windows.
        Breathe first fresh air in weeks, sigh deep like your

beagle after rubbing his belly. This morning your
    barn still stands. The goats and horses are hungry.
        Neighbors bring eggs, and for a moment

you are goldenbanner on the prairie, thunder of buffalo
    on the grasslands, bluebonnet sky—wide open
        above the deep-blood kettle.

The Lighthouse Keeper's Daughter

slips outside
her dreams

jolts awake
to bee-hum of surf

scratch of animal
on the roof

rips off
her eye mask

practices breathing
tabulates worries

before bed
she checked

the smoke alarm
and fire extinguisher

now she checks the
window latches

unplugs the space heater
googles

tsunami evacuation routes
on her phone

listens for the sound
of entering

when a blue-kite sky
doesn't calm her

she thinks about
her father cleaning

the Fresnel lens
on St. George Reef

how he lived on that rock
six miles offshore

in a tower
of granite and concrete

and how in the long howl
of a giant storm

the Pacific broke at the top
of the light

and shook with *ominous*
deep-throated rumblings

and when a stone carried up
from one of those waves

broke a window in the lens room
sending a surge of seawater

down the spiral staircase
somehow that refracted

light
kept shining

Driving over the Columbia River

Crossing the I-5 bridge
after calling the ambulance

for my father,
I think about borders.

The crossover where one place
becomes another.

How in the middle of the span
I both enter and leave.

How the place I call home
is nowhere and everywhere.

How the gyres of this great river
turn and turn—how she swallows

blossom and boulder
with her wide tongue of time.

Lifeblood

When asked
      what she will miss most

           she answers

                    all   that   water

boom of surf at Bastendorff Beach
    field of whitecaps on the Coos Bay Bar
        seasick swells of the Pacific

brisk current of Rosario Strait
    narrow boil of Deception Pass
        starlight twinkle of Admiralty Inlet

mirror of Mats Mats Bay
    foamy wake off Fidalgo
        sway of kelp beds off Burrows Island

when asked
    what she will miss most

        she knows

                  all   that   water

Tide Change

Write the names of those you've lost
in the sand of an Oregon beach.

Wait as long as the Moon requires
for the husk of crab and the polished shell

to pool
around your feet.

Some Aid to Navigation

The love-clock in a heart
        can take years to stop ticking.
                Each minute, day, year,
a slow winding down.

Some stop
        in the course of an hour
                as the world rotates
into night.

How different our lives might have been
        if somehow we had kneeled on the floor
                of our brokenness
listened to the howl of the wind at the door

and by memory and hope
        built a new kind of time.
                Then again, maybe it wasn't a clock
we needed—but a compass.

A tool to get our bearings, find coordinates,
        plot our course to where
                scanning the distance we could see
the faintest flicker of light.

III

*There is still a way to make your body a boat.
There is still a way to move through your life as
both water and boat,
no longer divided,
voice saying only "River, River."*

—Annie Lighthart from *Iron String*

Siren Song

As a child I heard stories
of the Keeper's Quarters
falling into the sea.

This did not stop me
from playing beneath the light
or following the pull of the cliff's edge.

The beach in front
of our old house is gone, the
ocean laps at its door.

Still, if given the chance,
I would build a house near
sandpiper and tsunami.

Choose water
even if
it drowns me.

The Lighthouse Keeper's Daughter Calms Herself Down

Let's pretend you're a girl again.
        Lie on the beach in a sleeping bag.
                Listen to the surf.

Rein your horse over remote logging roads.
        Never think: bear, cougar, man,
                foot caught in stirrup.

Wrap the baby crow in your shirt,
        drop sugar water into her beak.
                Suck crab claws like candy.

Fall asleep to the drone of foghorn.
        Burn your feet and have them buttered.
                Let an earache find a lap and warm towel.

Remember all the boats
        that carried you out and back,
                and how you stood on the bow.

Surfacing

to have even
an ocean's chance

of swimming
in that phosphorescent glow

you know
is your center

the one that too often
slips out between your legs

like silky kelp
or the men you love

but cannot hold onto
attach yourself

like a barnacle
to the belly of a whale

dive deep
into saltwater

place your finger
in the heartless

center
of a sea anemone

feel it grasp your promise
and beating pulse

embrace
its softening hold

Ode to My Belly

O belly,
little belly,

you entered
the world

a perfect
polished stone.

Do you remember
how your mother washed

the grapey globe of you?
Breathed the powdery slope

of you? Wrapped you
in her softest cotton?

How she pressed her face
against the dried raisin

of your connection?
How in the darkness of

those first long nights
she became bewitched

by the bubbly
yeast of you?

Garibaldi

At the Garibaldi Marina, The Spot fish market boasts
Fresh Live Local Seafood.

Buoys hang from soffits like tattooed breasts. Crab rings
stack like vertebrae.

The boats tied to the docks look just as I remember them,
only tricked out with more electronics.

At the edge of the parking lot, my father's
old 36-footer is dry-docked.

Here, no one remembers my Surfman father,
or the days he pulled bodies from the ocean onto this boat.

Here, no memory remains of men landing by helicopter to
pin a medal on his uniform.

Here, no one cares how many stormy nights he crossed
the Tillamook Bar, his hands icy cold on the wheel.

Here, even I am losing his story. There were so many boats.
So many rescues. So many things I got wrong.

Ode to Being Fed

O
Daddy—

my Kansas
winter.

My handheld
firecracker.

My Édith Piaf song
of no regrets.

My honey-butter
balm.

My small ocean
of fast boats.

My salty
tongue.

My long
highway

of road-rash
love.

Tonight, home
from the hospital

with your
patched-up heart

and pills,
you begin to cry

as I
microwave your dinner,

overcome by
the purple tulips

I have placed
on your counter.

Where God Resides

At the top of Neahkahnie Mountain
I release us—ask the south wind

to carry our burdens,
beyond lingcod and dogfish.

I stopped scanning the bar for
my father's boat years ago,

but on the dock I still pull crab pots
filled with his voice,

sometimes his face, once
the Pepé Le Pew tattoo on his calf

he tried to keep hidden from me.
I was so sure of who left who,

but this old coastal fog has blurred
things, left me alone with our bowed legs

and broad shoulders, a book about
a horse he gave me at fifteen,

chaps from that elk he hauled
off this mountain.

What else could I have done
but return to this western edge

we were our best in? To the dark
of Sitka spruce, where, in the smallest

shaft of light, we were sword fern unfurling
toward the blue-gray light.

Sometimes a Meteorite Becomes a Moon

On my daughter's phone
the mailbox is full.

When she finally calls,
I complain how hard it is

to leave a message.
She tells me to text,

says *don't take it
personally.*

Which leaves me
with the truth of it.

I miss
her voice.

        Funny thing
        about grief,

how it circles back
from the past into now.

How an unanswered
phone call

finds
the child in me

waiting
by a rusted mailbox

for my father's
postcards.

      The trick
        to remember

is that
hand-me-down grief

can be a comet
that curves

and its gold
shimmery dress

just a paper-doll
flash

while the Moon
o the Moon

still sparkles
on your

daughter's
dark hair.

Praise the Childhood of Borderless Roaming

Who can say
what catches the eye?

It might have been
the golden light in my hair

or the sheen of moisture
on my forehead.

It might have been the slow
pump of my legs

on her backyard swing, or the firs
fading from green to black.

But there she was,
slight frown on her face,

handing me a peach
the color of sunrise.

Looking back, she seemed
like a woman who had

seen things—
and held those things

deep within
the soft pillow of her belly.

And me, just a kid who knew nothing
of the heat that blankets

a valley, or the bordello fragrance
of trees laden with treacly fruit.

Just a kid she didn't know,
from a place of canned plums

and Dungeness crab, who can still feel
the heft of her offering,

its fuzzy skin on my lips,
it's Rubenesque plumpness,

creek of syrup
down throat and chin.

A kid who ate her
way to its red-edged center,

then suckled and suckled
on its dimpled slick pit.

Ode to the Cooper's Hawk

Nobody wants to talk about the breakdown of bodies.
The way finger joints become tulip bulbs beneath frozen
ground. How a thyroid can burn itself out like a star or a
spine can collapse like a sinkhole.

Drinking coffee at the kitchen table, I thumb through seed
catalogs, pray my winding road of a back bends to zinnias
one more season. Catch a movement
at the bird feeder.

Above the suet basket yellow talons grasp the shepherd's
hook. A large head over broad shoulders scans the dormant
lawn, the dark place beneath the ceanothus. Tail pointed
skyward she dives, wings tucked over her back.

A meteorite of feathers—that even after plummeting,
twists, then lifts, into the green-gold light.

Towards Kintsugi

Mother, do you remember the Smith-Corona
typewriter you bought in North Bend?

The one you carried in a hard case
filled with extra ribbon and carbon paper?

The one you typed your first college paper on?
The one I now have that won't type the letter *B*,

which makes the words heartbreak, or break,
or breaking, fall apart—like all my yellow anger.

Which might not be anger at all—but me
after school on an oyster road, mesmerized

by yellow daffodils wilding the yard of an
old house. And where under a sturgeon

sky I pulled stem after sticky stem, then ran
home where you took them from my hand,

placed them in a jar, and never once asked
where they came from.

Could it be my yellow anger
is not this broken typewriter at all

but the crooked line we took, to meld our
fractured parts into some kind of beauty.

It Came to This

My mother stayed put
the rest of her life

in those hills
where my father left us.

My father went back
to water.

I fled old rooms where
the briny smell of us

still hung in the air.
I drifted.

From boy to man,
from house to home,

from city to suburb.
Shrouded myself

with mortgage and
motherhood.

Closed the curtains
to the pull of the Moon.

There were times I was angry
at both of my parents.

Sometimes I felt erased.
Sometimes like the guest

no one knows in the kitchen
at a party.

I don't know what changed
really.

Perhaps my aging body.
Perhaps theirs.

But somehow before the end,
a door opened and we crossed over

our muddy thresholds. Somehow,
we served our finest ginger tea.

Birthday Fires

—with a line from Henri Cole

Bound tight to my mother, I am sung
into the world, around fire.

I am one of many kinsfolk
gathered on this beach,

one of many candles
glowing on the cake.

> So many birthdays before I know—
> *I came from a place with a hole in it.*

A hole deep enough to hold decaying marriages,
small harbors with drowning children,

scoliosis spines, sacrament of cheap beer, soft,
hard, dirty hands, placed against our cheeks.

Deep enough to bury
us all. Which is not to say,

we didn't drink glasses
of Asti Spumante. Salute the sky in Pisa.

Kiss behind tavernas—fiery candelabras
tilted on our heads.

The Lighthouse Keeper's Daughter Goes Home

Let's end with her palms
shell-pink and cupped,

scooping the waters
of the Little Nestucca,

Coos, Coquille. Here is
a woman who drags

her joy and sorrow
through lazy currents.

Here is a woman
who holds her loves

in old hands.
A woman who patiently

waits, until they meander
or rush back

to the end
of these river runs—

to the sound
of the ocean's

open
mouth.

Notes

"Birth Announcement": Nellie's Cove is below the Port Orford Lifeboat Station in Port Orford, Oregon. In 1955 when my family lived here, it was accessible from the station by a staircase of 532 steep wood and concrete steps. The cove held the boathouse which housed two 36-foot lifeboats.

"The Coxswain": A coxswain in the United States Coast Guard is the person in charge of a small boat. The coxswain has the authority to direct all boat and crew activities during the mission and modify planned missions to provide for the safety of the boat and crew. This poem was inspired by Ada Limón's "The Raincoat."

"The Lighthouse Keeper's Daughter": Words in italics are quotes from Floyd Shelton in the book *Sentinel of the Seas* by Dennis M. Powers, p. 300. Powers interviewed my father about his time living and working on St. George Reef Lighthouse in 1952, which included his description of what it sounded like inside the light during the "big storm" of 1952.

"We're Heading into Nut Country Today": Words attributed to President John F. Kennedy spoken to his wife, Jackie, on the morning of November 22, 1963, after showing her an ad the John Birch Society had placed in the *Dallas Morning News*, alleging that the Kennedys were pro-communist.

Acknowledgements

Many thanks to the editors of the following journals in which some of these poems first appeared, sometimes in slightly different forms or with different titles:

Cirque: "Returning," "Closing Time"

Dodging the Rain: "Surfacing," "The Days of Starfish"

Empty House Press: "Praise the Childhood of Borderless Roaming"

Kitchen Table Quarterly: "War Play" (previously "After Ukraine is Invaded")

Muleskinner Journal: "Where God Resides"

Neahkahnie Mountain Poetry Prize: "Birthday Fires"

North Coast Squid: "Ode to the Cooper's Hawk"

Shark Reef: "Early Seral Stage," "Birth Announcement"

Snapdragon: "Solastalgia"

Tabi Po: "Driving over the Columbia River"

Verseweavers: "Extinction Dreams after Beachie Creek Fire"

Wild Roof Journal: "Ode to Lust" (previously "Ode to the Satellite")

Gratitude

Deep gratitude for the Sou'wester and Hypatia-in-the-Woods artist residencies which gave me uninterrupted time and support to bring this collection to fruition.

A big thank-you to poetry instructors Gary Lilley and Emily Ransdell whose workshops and classes have been instrumental in my becoming a better poet.

And to all the poets in all the workshops I have taken, who either listened to or provided feedback on some of the poems in this manuscript, thank you. And especially to those of you who have become my friends in the process: Gina Hietpas, Stephanie Nead, Charlene Bushnell, Anne Murphy, Carole Anne Modena, Gary Bullock, Dan Coffman, and Connie Soper.

With gratitude to Mary O'Shaughnessy who was gracious and responsive in allowing permission to use her art as the cover for this book. Writing this book took me through a portal to my past, while you, someone I did not know, had created an image of a portal to an aid to navigation known as Point Wilson Lighthouse. The first lighthouse I lived at. Thank you.

To Theresa Page who believed in me and provided me with a safe place to understand my story.

To Lana Hechtman Ayers, MoonPath Press editor who I first met online during Covid in a poetry class. Her support of my work has made this journey of publication a joy and a pleasure. And to Tonya Namura, book designer, for her mastery in creating an aesthetic home for these poems.

Love to my parents. Thank you for sharing yourselves with me and for understanding my need to process our shared life.

And finally, to my husband Charles, who never shied away from my grief. You gave me the foothold I needed to share my truth.

About the Author

Carey Taylor was born in Bandon, Oregon, shortly after her parents moved into their first home, next door to the Port Orford Lifeboat Station. She grew up following her father's Coast Guard career on the Oregon and Washington coasts, and had the good fortune to live at Point Wilson Lighthouse, Burrows Island Lighthouse, and Cape Arago Lighthouse as a child.

Some of her first memories are her view of the Garibaldi Boat House from her bedroom window at the Tillamook Coast Guard Station, helicopters landing in her front yard, and the persistent background wail of a foghorn. Her love of this western edge of the world has been an integral part of her identity and one of her greatest writing muses.

Carey is the author of *The Lure of Impermanence* (Cirque Press). She is the winner of the 2022 Neahkahnie Mountain Poetry Prize, a Pushcart Prize nominee, and runner-up for the Concrete Wolf Louis Poetry Book Award. She has been published in the United States, Ireland, and England. She has a Master's Degree in School Counseling from Pacific Lutheran University.

Carey lives in Portland, Oregon. You can visit her website careyleetaylor.com.

Printed in the USA
CPSIA information can be obtained
at www.ICGtesting.com
CBHW031641150824
13252CB00015B/698